ANIMALS EXPOSED!

The Truth About
Animal Intelligence

Created and produced by Firecrest Books Ltd
in association with the
John Francis Studio/Bernard Thornton Artists

Published by Tangerine Press,
an imprint of Scholastic Inc;
557 Broadway, New York, NY 10012

Tangerine Press and associated logo and design
are trademarks of Scholastic Inc.

ISBN 0-439-51808-3

Printed and bound in Thailand
First Printing December 2002

The Truth About
Animal Intelligence

Bernard Stonehouse
and Esther Bertram

Illustrated by
John Francis

Tangerine Press and associated logo
and design are trademarks of Scholastic Inc.

FOR MATTHEW

Art and Editorial Direction by
Peter Sackett and Dr Bob Close

Edited by
Norman Barrett

Designed by
Paul Richards, Designers & Partners

Color separation by
**SC (Sang Choy) International Pte Ltd
Singapore**

Printed and bound by
Sirivatana Interprint Public Co., Ltd., Thailand

Contents

Introduction

Most of us can sympathize with animals. In their daily lives they meet problems and successes that we can recognize. An ant carrying a seed twice its own size, a young bird landing dangerously from its first flight, a cat carrying its kittens to safety – we sympathize as we would if they were humans in trouble. We tell stories in which animals behave like people, some true, others simply stories or myths. It is hard to tell one from the other.

Are animals like humans? Only up to a point. Human behavior starts in our brains and nervous systems. Animal behavior starts in many different kinds of brains. Jellyfish, worm, snail, and crab behavior starts in much simpler brains and nervous systems, yielding simpler patterns of behavior that get those animals by in their simpler worlds. Fishes, amphibians, reptiles, and birds are more complex. Mammals, the most complex of all, come closest to humans, with primates (monkeys and apes) topping the list.

Animals and humans come closest in behavior that is based on instincts, the patterns that both we and they are born with. To these simple patterns the more complex animals add learning – remembering and thinking things out in ways that the simpler animals cannot. There is not much you can teach a jellyfish, but dogs, elephants, and people can learn all sorts of things very quickly. Instinct and learning together serve us well. This book shows how they combine to produce behavior in animals from the simplest to the most complex – including ourselves.

Simple animals, simple senses

Blowfly larvae respond to light and humidity, but to very little else that goes on around them.

It is often thought that animals behave much as we do. The truth, however, is that animals of various levels of complexity have different and usually lesser abilities than humans. Only primates approach our level. For example, these maggots, the larvae of a blowfly, live very simple lives. The mother blowfly, strongly attracted by scent to rotting meat, laid her eggs on a recently killed rat. The eggs hatched into tiny maggots, which have only one thing to do in life – feed on the meat. They live in a film of moisture, staying out of strong light, and moving a fraction of an inch if the surface becomes dry. For this they need only simple senses. They are growing quickly, and each will develop into a pupa before emerging as a blowfly.

On these pages are an amoeba, one of the simplest animals we know, and a young chimpanzee, one of the most complex. The jellyfish and oyster represent two of the millions of different kinds of animals that lie between these extremes.

Drifters
Dozens of different kinds of jellyfish float close to the ocean surface. They move by pulsating – a muscular movement of the "bell" that forces water downward and keeps them afloat. With only very simple sensory organs, they can shy away from strong light. Their tentacles are sensitive, moving to entangle and capture fish that touch them.

Simplest animals
Amoebas are some of the simplest animals we know. The size of a pinpoint, each water-dwelling creature is made up of a single cell

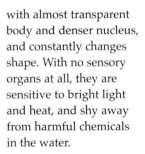

with almost transparent body and denser nucleus, and constantly changes shape. With no sensory organs at all, they are sensitive to bright light and heat, and shy away from harmful chemicals in the water.

Stuck on the rocks
Oysters are fixed to the rocks on which they live. Unable to move away if danger threatens, they can only close their shells. Tiny light-sensitive eyes tell them if birds or other predators are near. They also close up if the tide falls.

Almost human
Though born small and helpless like a human baby, a young chimpanzee is fully equipped to see, smell, hear, touch, and feel the world around it. While still with its mother, it starts to experiment, and it learns by trying new things. Monkeys and apes (which include chimpanzees) are the only animals that approach human abilities.

Born ready

Are young animals relatively helpless? It may seem so, but all are equipped in some way to get by from the moment of birth. This wildebeest calf lives on the dry, grassy plains of Africa among a herd of thousands. Early this morning, its mother gave birth. Within minutes, it was standing upright on spindly legs and nuzzling for its first drink of milk. As she moved, the calf took its first steps. Within half an hour, it was walking unsteadily. Now the herd is resting. Soon they will move on, and the calf will be strong enough to trot along with them.

Like other animals on this page, the newborn calf has just enough senses and just enough strength to see it through its first few difficult hours of life. Protected by its mother, and by the rest of the herd, it will learn quickly. It stands a good chance of surviving.

Basic swimmer
Only a few minutes ago, these tiny tadpoles broke out of the eggs in which they were born. So far they cannot do much, except straighten their bodies and contract the muscles alternately on either side. That sets their long tails wiggling and pushes them up to the surface of the water, where they can gulp their first air bubbles.

Wildebeest (also called gnus) are large antelopes with strongly curved horns. They move constantly in search of new grazing lands.

One kind of move
This baby leatherback turtle, matchbox-sized and only an hour old, moves its front and back flippers alternately like a clockwork toy. That movement was all it needed to climb from the sandy pit nest and race down the beach toward the sound of the breaking waves. Now it will swim off into deep water.

Hatchling: one response
This baby bullfinch broke out of its shell yesterday. Though still blind, today it can make the one movement it needs to survive. When it feels the nest shake, it shoots up its head and gapes with bill open, so the visiting parent can stuff food into its mouth.

Learning to be a duck
This day-old duckling hatched from its egg all ready to go. Complete with a downy covering, already it can walk, see, hear, and make cheeping noises. Its first tottering steps followed the first moving object it saw – its own mother. Soon it will swim with her and learn to dabble for food. In a few months' time it will seek out a mate that looks just like her.

Skills for catching food

Sight and movement
Pond skaters skimming over the water use their large compound eyes to look for the smaller insects on which they feed. Sensitive also to tiny ripples, they skate over to investigate any small animal that settles on the water nearby. They are stimulated to attack by the little insect's movements.

Group hunting
Lions are more effective in bringing down large prey when they hunt together. Living together teaches them to watch each other, and spreads skills among them. These lionesses, creeping up on a group of antelopes, have worked together many times before. Each knows the tricks of lying low, moving slowly, and waiting for the right moment to attack.

Does hunting involve skill and cunning? For lower animals, such as the sea anemone and the pond skater, hunting is a simple action involving very little skill. For insects such as the hunting wasp, it can become complex, though it is still made up of simple elements.

Only in birds and mammals does hunting clearly involve memory and learning. This burrowing owl of western North America has caught a large grasshopper. A skilled flier, the owl may well have caught it in the air, holding it with its strong curved talons. Tomorrow it could just as efficiently catch mice or rats on the ground, or knock small birds from the branches of trees. Nobody taught the owl these tricks; older animals do not teach younger ones how to hunt. It might have learned by watching its parents, but most of its skills came from trying, succeeding, and remembering. Lions learn hunting skills, also, mainly by trial and error.

Hunting the easy way
Settled in a weak current on the seabed, all the sea anemone has to do is spread its tentacles. The current brings in tiny particles of food that get caught on the tentacles, and the anemone pulls them toward its central mouth. Occasionally it loosens its grip on the rocks and moves to another site.

Sight and touch
This hunting wasp lives by catching bumblebees. Some it kills and eats immediately, others it stings and carries to its nest for its larvae to feed on. How does it recognize bumblebees? It chases and grabs any flying insect that is about the right size, but it kills only the ones that feel fuzzy like a bee.

Burrowing owls hunt by day or night. They live in burrows that they dig for themselves or take over from other animals.

Animal memory

Forgetful robin?

Robins build nests in hollows. This robin started to nest in a stack of pipes, bringing dry moss to one of them, then to a second and third. He ended with one nest finished and several dozen half-built. Was he forgetful? Nest-building is not so simple as it seems.

Long memories

Elephants trek hundreds of miles across dry African plains in search of water and food. Usually a troop follows a circuit that takes them to a succession of likely places, where water and trees with fresh leaves have been found before. In this they seem to rely on the memory and experience of older members of the troop.

These macaroni penguins have only a few weeks of good weather for raising their chicks.

These macaroni penguins (opposite page) spent the winter several hundred miles away at sea, where they find their food. Now, in spring, they have returned to their nesting grounds in sub-Antarctic South Georgia Island, where the air temperature is below freezing and snow is still on the ground. Each penguin returns to the exact site where it nested the previous year. How does it find its nesting spot? We do not know. Many kinds of birds that find their way home over vast distances use the sun or stars to navigate, then home in on local features. Young penguins often return after two or three years of travel to nest in the colony where they were hatched, then continue to return to the same colony each year.

The longer animals live, the more useful their memories become. Elephant troops rely on the memories of their older members to find food and water. But memory also works for insects and small birds with much shorter lives. Do animals remember as we do? Different animals remember different things that are important to them. But even elephants forget.

Short-term memory

Can bumblebees remember? They often make repeated visits to a cluster of nectar-laden flowers several hundred yards from their nest. This means they remember where the flowers are and what they look like. They know how to navigate between the flowers and the nest.

Forgetful jays?

In late summer and fall, when berries and other fruit is plentiful, jays eat as many as they can. When the jays are no longer hungry, they continue to hunt for them, but hide them in holes in tree trunks and grass. In winter, when food is scarce, they find and eat some of the berries they hid earlier. However, they probably find less than half. They forget where they hid the rest.

Defending the home

Fighting a rival off

Cockerels of domestic poultry, like similar birds in the wild, attract a group of hens around them, and keep rivals away. Each group moves continually around a large territory, scratching the ground for insects and seeds. The cock crows from time to time to warn rivals. If a rival cock challenges him, the two fight, striking with

bills and sharp claws. The winning cockerel takes over the group of hens, and the loser wanders off to start another group.

When we think of animals defending their homes against predators, we may think of human homes. But home means different things to different animals. For some it is a den or nest, for others a territory – a piece of land where they live and feed. Either way, they spend a lot of time and energy trying to keep rivals away. This great skua (opposite page) has a nest on the ground by a group of rocks, where his mate is taking her turn to incubate two eggs. His territory is an area about 1.25 miles (2 km) all around, where he and his mate hunt for most of their food. They take little notice of gulls or other birds crossing the territory, but they raise their wings and call loudly when another great skua flies over. This display warns other skuas to keep away, or to expect trouble if they try to land. If the visiting skua flies close to the nest, the territory owners take off and attack. Other animals defend their homes or their mates in different ways, usually with a warning first, then by action.

Buzz and sting

Black-and-yellow tree wasps live together in groups of several thousand, centered around an egg-laying queen. Wasps flying in and out of the nest attack any large animals that come too close, first buzzing them in warning, than landing on them and stinging. One sting is painful. Several together can kill.

Keeping them guessing

Blennies are small fish that live in rock pools and lay their eggs in cavities among the rocks. A blenny with just one nest may defend several cavities in the pool. So a predator passing by can never be sure which cavity contains eggs.

Winner takes all

These large white Dall's sheep live in the mountains of the northern United States and Canada. In the fall, rams claim hillside territories and gather small groups of ewes about them. The groups keep out of each other's way. But should a rival ram approach a group, the "owner" ram squares up to him, usually chasing him off. If the rival persists, the two clash horns, butting and pushing each other in a noisy show of strength. The winner stays with the group, mating with all the ewes.

There is no mistaking a great skua's threat – the spread wings with prominent white bars and the long, raucous call tell visiting skuas to keep away.

Sleeping or hibernation?

Disappearing tortoise

If you keep a pet tortoise, you may find it disappears during the fall. Like the bear, it stores up fat, then finds a safe corner where it can sleep undisturbed. In a very cold winter, it may be safer to keep it in a box in a cool, dry basement.

Do bears hibernate through the winter? Many think so, but that is not strictly true. Some animals, like this brown bear, sleep solidly for several months each year, but they are not truly hibernating. They are big animals and need a lot of food to keep them going. They eat fish, carrion (dead animals), berries, eggs, and all sorts of other foods, so there is plenty for them in spring, summer, and fall, but little in winter. So they fatten during the fall, then find a comfortable cave or den where they can curl up and sleep for weeks on end. This means they use little energy, just enough to keep them alive and warm. In contrast to the bear's deep sleep, true hibernation is when the

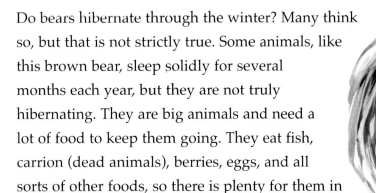

Mouse that sleeps

"Dormouse" means "mouse that sleeps." A well-known hibernator, it wraps itself in grass and leaves, in a burrow or hollow, or sometimes even in old clothing inside people's houses. It curls up tightly in a ball and sleeps for four or five months. Upon waking, it takes several hours to recover normal body temperature. It can be harmful to wake it before the winter is over because it will be very weak.

Safe in the snow

European hedgehogs spend much of their winters in hibernation. They find a place where they can curl up, well-insulated from the cold, and drop into a deep sleep. If it snows, the snow piles up around them, acting like an extra quilt to keep them warm. During this sleep, which may last several months, they lie tightly curled up with their spines erect to keep predators away.

Autumn feeding

For a bear to survive the winter, it has to feed well in the fall. Rich, fat-filled salmon can be found in the rivers at just the right time of year. Bears can stock up and sleep undisturbed, using just enough fat every day to keep themselves until spring.

heart rate slows and the body temperature falls to that of the outside world. A number of animals remain in this state during the winter, including dormice, tortoises, and hedgehogs. If you see them, don't wake them up, so they can save their energy to survive the long winter.

Opposite page: A brown bear sleeps, undisturbed in its cozy den during the winter.

In close quarters

Cooperative effort?
Coral reefs are the work of millions of tiny creatures living in close quarters. There is no cooperation between them. Each animal is like

a miniature sea anemone, looking after itself by catching food in tiny tentacles, and ringing itself with a tube of hard chalky material in which it can live safely.

Social insects
Termites are social insects. Some live in timber, others in soil, but best known are the mound termites of southern Africa and Australia, which build nests towering up to 13 feet (4 m). The thousands of termites in each mound are all worker offspring of a king and queen that live in the center.

Storybooks often make animals out to be friendly and warm toward each other, particularly when they live in close quarters. This is not usually true. Animals that are hunted have to be wary of predators, so they like to keep space around themselves. Herd animals that live close together keep their distance from each other while grazing or moving. These Mexican long-nosed bats roost together in a mountain cave (opposite page). They hang from the roof, finding comfort in each other's company, but squabbling constantly among themselves and snapping at each new arrival.

Do animals cooperate to build shared homes? The truth, most likely, is that individuals usually work individually to produce the overall result. Wasps, termites, and meerkats have all worked out alternative ways of living in close quarters. Coral reefs, made up of millions of tiny individual animals, represent real crowding.

Living in burrows
Meerkats, mongoose-like animals up to about 14 inches (35 cm) long, live on the plains of southern Africa. Families of 20 to 30 adults and young live together in burrows and passages that wind through the soil, with several openings to the surface. They feed on insects, spiders, lizards, snakes, and other small animals, hunting within a few yards of the burrow entrances. Living together adds to their safety. Alert to hawks and other predators, they bark when alarmed, warning each other of danger.

Crowded quarters
Tree wasps build delicate, papery nests of chewed-up wood that hang from tree branches. The queen starts the nest, and her worker offspring later add to it; 6,000 or more wasps may

eventually live together in a nest the size of a basketball. The workers forage for nectar and insects, bringing home food for the growing larvae.

Opposite page: Mexican long-nosed bats live in groups but quarrel constantly. This newcomer cannot expect a warm welcome and must find its own spot on the cave roof.

Keeping the peace

Colony of kings
King penguins, like fur seals, breed on cold southern oceanic islands, forming colonies numbering tens of thousands. Each pair raises a single chick, which takes about a year to mature. Therefore, chicks must remain in the colony throughout

winter. Though the parents space themselves out, the chicks, in dense brown down, huddle tightly together to keep each other warm in the below-freezing temperatures.

Humans are sociable animals, living together in crowded towns and cities. We may get the idea that animal groups such as seal and penguin colonies – even beehives and termite nests – work in similar ways. Not so. Different kinds of animals have different reasons for crowding together. Most are to help the species survive.

These fur seals on South Georgia, an island in the cold South Atlantic Ocean, come together for breeding in spring and early summer. Around October, each large male takes up a small territory of a few square yards where the pregnant females will come ashore. When the females arrive and give birth to their single pups, the resident male takes charge of the females in his territory, fighting off other males. The mothers feed their pups until March, when all the seals disappear to sea, leaving the breeding colony empty.

King penguins and kittiwakes come together in the thousands for breeding. Killer whales live year-round in family groups, and honey bees have their own special recipe for communal living.

Bees in a hive
A beehive is a man-made house in which honey bees live as naturally as possible. Their life centers around the queen, a female that lays up to 5,000 eggs per day in wax cells made by the worker bees. Workers gather pollen and nectar to feed the larvae.

Clifftop dwellings
Kittiwakes are small gray gulls that nest in huge colonies on cliffs overlooking the northern oceans. On the crowded rock ledges, each pair has a tiny nest site of a few square inches, just big enough for a bird to incubate two eggs, and for two chicks to grow up side by side. Parents leave the nests to feed on small fish and plankton. Why nest on cliffs? To be safe from foxes and other ground predators.

Group hunting

Killer whales, or orcas, hunt together in family-based groups, usually made up of several mature females with their calves and some of their young from previous years. Different groups of orcas hunt in different ways, some specializing in catching fish, others hunting mainly seals or other kinds of whales. Hunts are well organized. Individuals call to each other, seeming to work cooperatively like lions.

Each mature male fur seal gathers around him a dozen or more females, mating with them soon after they have pupped, and keeping other males away.

Showing off

Ecstatic penguin

Penguins don't sing, but they call and display at breeding time. It is early spring, and this cock adelie penguin has returned to his nest site in Antarctica. He has marked the nest with stones, and gives an "ecstatic call," stretching up with bill in the air. This will attract any unattached hen penguin nearby – possibly his mate from last year, possibly another if his mate is late in returning.

Bird song

Do birds sing because they are happy? No – song is a cock bird's way of announcing his presence. This cock chaffinch took up his territory in early March, singing mainly in the early mornings and evenings. He found a mate, but continues to sing, telling other chaffinches that he is still around, and keeping them off his patch. His first song was very simple. Now he has a more elaborate one, learned from listening to other chaffinches.

Animals displaying to attract a mate are among the splendors of nature. Do they enjoy showing off? Perhaps not, because there could be hidden dangers. They may also attract predators. For insects, birds, and most small animals, springtime is mating time, so that offspring are produced in summer. Larger mammals must mate earlier to allow spring births. Attracting a mate usually involves showmanship, often the male giving a display that stimulates the female to accept him as a partner. Birds give some of the most vivid displays.

Here a cock argus pheasant spreads brilliant plumage and dances before a drab but interested hen. Pheasants, frogs, and chaffinches all endanger themselves to find a mate. If a predator came by, it would be attracted to bright colors or loud calls. But tigers have little to fear by attracting attention. For the others, it seems to be worthwhile, if only for the few weeks of the breeding season.

Right: After displaying, the cock argus pheasant is often able to mate immediately with the hen. If not, he displays constantly until she accepts him or walks away.

Frog song

Frogs are sometimes brightly colored, and color may be involved in their courtship. However, males usually court in the water, relying on sound to announce their presence and attract females. Loud and persistent croaking announces that a male frog has found a patch of water and is waiting for a female to call back. But it may also invite a predator.

Little to fear

Tigers live mainly in dense forest, among trees and dense undergrowth. If a tiger wants his presence to be known, he must make a loud noise that carries through the undergrowth. He has little to fear from others hearing his low, far-reaching roar.

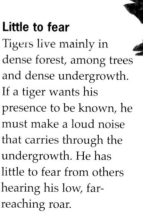

Looking for a home

Tree hollows
Like the closely related North American chickadees, European blue tits seeking nest sites look for hollows in old trees. Tiny under their dense covering of feathers, blue tits can squeeze through a hole

We may believe that animals look around for comfortable homes. But the truth is their instincts provide the impetus for what often result in strange but practical results. The overriding instinct of every animal is to produce more of its kind, and for many animals that involves finding a safe home. For some, the home needs a territory around it. (See, for example, the great skua, page 17.) For others, the site itself is all that matters. European robins like a covered nest site close to the ground. This one (opposite page) seems well-settled in an old flowerpot. That would not suit the birds on this page, the chickadee and house martin. Animals as different as hermit crabs and codlin moths also need homes and come up with quite different answers.

Red-breasted European robins are often prominent in gardens, especially during the hungry winter months, but hide their nests well away from sight.

Looking for a shell
Hermit crabs, marine crustaceans with curiously twisted bodies, live in discarded whelk shells. They start life as eggs, which develop into floating larvae. After drifting for several months in surface currents, the larvae descend to the seabed and find a small whelk shell, fitting inside with just the head, claws, and legs sticking out. As they grow, they need to find more whelk shells, each larger than the last.

less than 1 inch (2.5 cm) across, and raise a brood of a dozen chicks in a cavity 4 inches (10 cm) across. Where old trees are scarce, they take readily to nest boxes of these dimensions. Put up nest boxes in late summer, and the blue tits or chickadees will find them in time for spring breeding.

Under the eaves
Away from civilization, house martins nest on cliffs and crags. In towns and villages, eaves of houses are a welcome substitute. They build their nests of mud, carried by the beakful from nearby ponds and plastered into a sturdy cup under the roof.

Home is an apple
If you have ever bitten into an apple and found a gray grub inside, it was probably the larva of a codlin moth. To these small reddish-brown moths, home is an apple tree. In early summer, the moth lays its eggs on a

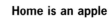

young apple. The eggs hatch and the larvae munch their way to the center, feeding on the fruit and pips. The larvae winter as cocoons in the bark of the tree, emerging as moths in spring.

Black-headed weavers of Africa often hang their nests from the tips of palm fronds, making it hard for snakes or rats to steal the eggs.

Pile of sticks

Ospreys need little skill in building their nests. Each spring, these sea eagles return to the same site. Their first action is to bring in sticks and twigs, making up for losses during winter storms, and adding a little to the pile. The hen osprey stands in the middle, arranging the twigs around her, then lays her clutch of three eggs.

Building a home

Surely nest-building calls for thought and skill, doesn't it? The truth, more often, is that it involves actions based on instinct rather than learning. Of all animal homes, bird nests are the most obvious – often neatly made, and strong enough to contain a parent and a growing family. Other animals, from spiders to cobras, also build nests in which their eggs and young can develop safely.

Among birds, weavers produce some of the most elaborate and beautiful nests. This cock black-headed weaver (opposite), just starting a nest, is knitting a ring of palm-frond strips. Soon it will become a hollow ball. At that stage, he will hang underneath it, flapping his wings and calling to attract a mate. When he finds one, she will line the nest and lay two or three eggs, which she will look after alone. Meanwhile, the cock will weave several other nests and attract more mates.

Other kinds of nest are simpler. Cobras gather a pile of leaves, pigeons a few twigs, and ospreys a pile of branches. The canary builds her nest in stages.

Simple nests

Some birds build no nest at all. King penguins (page 22), for example, hold the single egg on their feet. Cliff-nesting sea birds like kittiwakes (page 22) often get by with little or no nesting material. Rock doves, relatives of the pigeons you see in cities, nest on cliffs and crags in the wild, also using little material. City pigeons nesting on roofs and windowsills do the same. This one has gathered just enough sticks to keep its eggs from rolling.

Nest of snakes

Some snakes, lizards, and crocodiles make nests for their eggs. Though never so elaborate as the more complex bird nests, they keep the eggs together in one place, where one or both parents can look after them. This cobra has pulled together a pile of leaves, laid 20 white, soft-shelled eggs, and gathered more leaves over them. She doesn't incubate the eggs, but she'll stay nearby and threaten any animal that comes too close. The young snakes will never meet their mother. Upon hatching, they will slither quietly away.

Building in stages

A hen canary with eggs forming inside her, starts nest-building by collecting fine grass and laying it in the fork of a bush. She sits on it, trampling it under her feet and shaping a bowl around her, then collects more grass. Hormonal

Fine grass first

Shaping the bowl

Soft lining of feathers

(chemical) changes in her body cause her to pull feathers from her breast and abdomen. With these, she makes a soft lining. Once she has started to lay, the feel of the eggs against her body make her "broody," or ready to incubate.

Nursing and feeding

Elephant milk

Like all other mammals, elephants feed their young on milk. Baby elephants, born singly, find the milk in two mammary glands. The mother knows instinctively to protect her young, but she may learn how to do be a better mother as she has more babies.

Do animals care for their young as we do? They may appear to, but most know instinctively how to raise their young, unlike humans, who learn as they go or from other humans. These burying beetles are a good example of instinctive behavior. A male and female, each about 1 inch (2.5 cm) long, are burying a dead shrew, digging away the soil underneath to lower it into the ground.

When they are done, the female will dig a tunnel under the body, and lay a dozen or more eggs in it. The eggs will hatch into maggot-like larvae, which will shuffle up to the rotting carcass. The mother will feed them first on juices from the flesh, then leave them to feed themselves. They will grow fat, developing through several stages into hard-shelled pupae, from which new burying beetles will eventually emerge.

These beetles have little brain for learning or thinking things out. Wasps are mainly instinctive. Birds and mammals have instincts, but they also learn as they go along.

Right: Burying beetles, drawn by scent to a dead shrew, are about to bury it.

Family larder

Mason wasps are small flying insects, half an inch (1.25 cm) long. They dig tunnels in sandy walls and lay eggs, each in a separate chamber. Then they instinctively imprison live caterpillars in the chambers for the growing larvae to feed on.

Cropful of food

Many kinds of birds carry food home to their chicks in their crop. This hungry brown skua chick

is whistling and tapping its parent's bill. Soon the parent will vomit a cropful of fish onto the ground for the chick to eat.

Food for cubs

Wolf cubs grow quickly and soon need more milk than the mother can supply. Both parents go hunting, leaving the cubs on their own in a well-hidden den. Different seasons provide a wide range of prey, including reindeer, hares, rabbits, birds and their eggs, and mice. At first, the parents swallow and partly digest the food, throwing it up for the cubs on the floor of the den when they return. Later, they learn to bring back whole carcasses, which they share with the growing cubs.

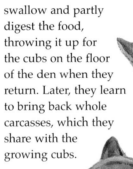

Pecking the ground

Some young birds need feeding by their parents for days or weeks. But this chick of a domestic hen, only a day old, is already feeding itself. How does it learn? Instinct tells it to peck the ground, first at random, then at small, loose bits that look different. Within hours, it learns to tell small stones from seeds.

Eating right

Tasting mother's choice

A baby koala from southern Australia lives first in its mother's pouch, feeding entirely on milk. As it grows, it leaves the pouch and clings to her back, where it can take an interest in the world around it. The young koala sees and smells the plants that its mother is eating, from time to time leaning over and tasting them too.

We are inclined to think that animals instinctively know what to eat. Is this true, or do they learn as they go? Certainly, the lower animals, such as insects and spiders, tend to have simple, instinctive diets. But with birds and mammals it is not so straightforward. There are dangers in feeding. What looks or smells good to a hungry youngster may be poisonous. Some young birds are programmed to select the right kinds of food, but most are fed initially by their parents. Young mammals feed first on milk, then on the goodies their parents bring home, watching and learning as they go. It is no coincidence that our noses and tongues lie close together at our food intake site. Both are sensitive chemical testers that we use in selecting good food from bad.

This young tamandua in the South American rainforest (opposite page) is learning how to catch and eat ants. Young chicks, gorillas, koalas, and grizzly bears also learn important tricks for feeding.

Learning from mother

Grizzly and brown bears are big animals with big appetites. Too heavy and lumbering to climb trees for bird nests, or chase deer and rabbits, they cannot afford to be choosy. They catch fish, but eat a mainly vegetarian diet of grasses, shoots, and berries. This cub will travel with its mother for well over a year, learning from her what is good to eat.

This tamandua cub watches as its mother follows an ant trail, sniffing with her sensitive nose and licking ants from the bough with her long, sticky tongue. Soon it will try for itself.

Learning from the group

This young gorilla moves through the central African forest with his family group. For months, he rode clinging to his mother. Now she has another baby, and this one is big enough to walk by himself. The group moves constantly, seeking roots, fresh shoots, buds, fruit, and other vegetable foods. The young gorilla watches to see what they eat, and he samples the different kinds of food with them. Ripe bananas are his favorite food.

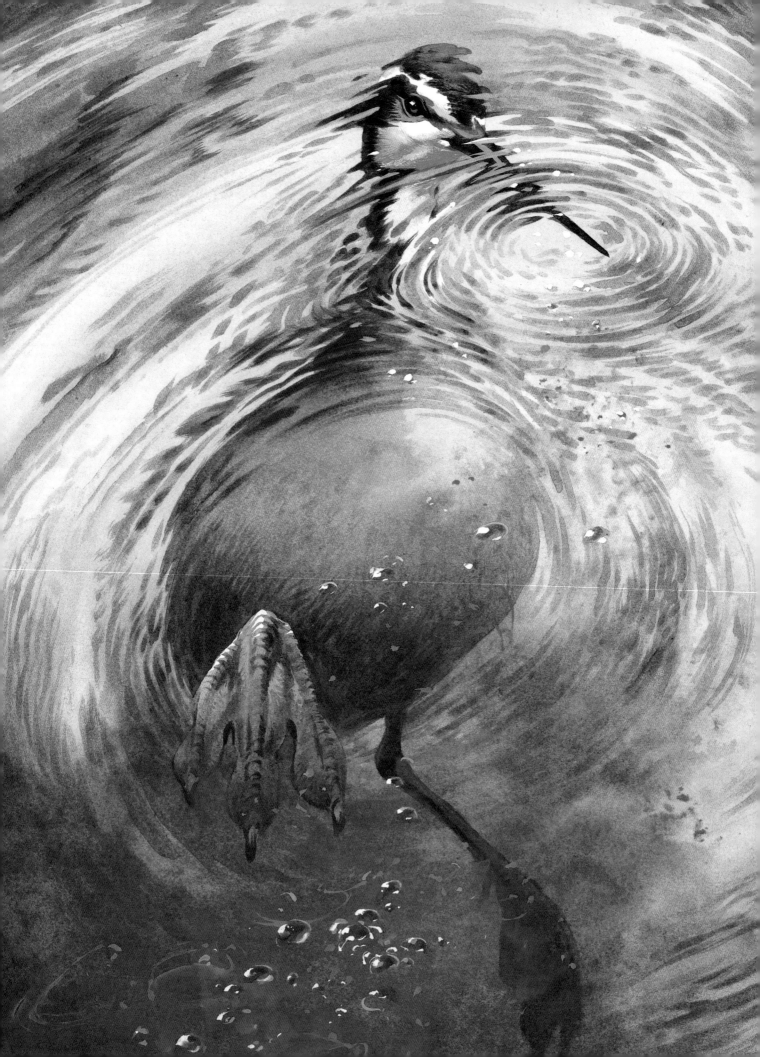

Hunting and trapping

Animals have found all kinds of ways to hunt and trap each other for food. Are they very clever and cunning? We may think so, but their success is often based largely or entirely on instinct, less often on learning or experience.

Wilson's phalaropes (opposite page) are birds about 10 inches (25 cm) long. They feed inshore in summer, but forage many miles from land in winter. They catch food by spinning like tops in the water, drawing floating particles of food toward them.

Humpback whales make curtains of bubbles to trap plankton and small fish. The polychaete worm spreads a net of tentacles, the pistol shrimp stuns fish with sound, and the spider spins webs to trap insects. These are entirely instinctive actions, though a spider's second and third webs are usually better than its first.

Spinning a web

Orb, or garden, spiders spin elaborate webs of fine silk. Starting with a square or triangle, the spider adds crossover cables that meet at a central hub, then lays down a spiral of sticky threads that become the main trap.

Trapping by tentacles

Polychaete worms, distantly related to earthworms, live in tubes on the seabed with just their heads sticking out. The tentacles, spread in a mat over several square inches, trap small fish or other animals and draw them toward the mouth.

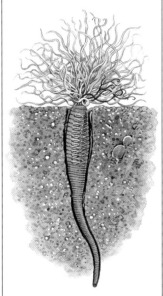

Stunning by sound

Pistol shrimps live in the sandy seabed off the east African coast. One claw is elongated and altered to form a hammer that can be cocked and fired like a gun, emitting a sharp click. The sound stuns tiny fish, which the shrimp grabs and kills.

Bubble curtain

Humpback whales, up to 60 feet (18 m) long, feed on plankton – the thin soup of tiny animals that float at the sea surface. To concentrate the plankton and make a thicker soup, one or two humpbacks dive deep and spiral up toward the surface, releasing a stream of bubbles from their blowholes (nostrils).

The bubbles rise in a cylindrical curtain, drawing the plankton together into a central mass, on which the humpbacks then feed.

Phalaropes feed by pecking at small surface particles. Spinning around may draw food within reach of their bills.

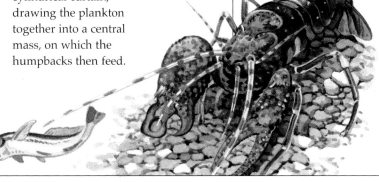

New tricks for feeding

Nut hunt

Squirrels love nuts. The end of summer, when many kinds of nuts are ripening on the trees, is the high point of their year. But squirrels were quick to learn that garden birdfeeders, put

out for titmice and other small birds, hold peanuts year-round. Special feeders are often needed to keep squirrels out.

Cracking crabs

Gulls scavenging along the shore sometimes catch crabs before they can scuttle for shelter. How does a gull break into a hard-shelled crab? Herring gulls take them up high and drop them onto hard rocks. Small crabs bounce and run off, but larger ones may be stunned and damaged enough to be split and eaten.

Can animals devise new ways of feeding they have not used before? Yes, up to a point, but new ways are most often variations on old ways, discovered by trial and error. Japanese macaques (a kind of monkey) live by the sea, feeding on crabs and other seafood they find along the shore. A few years ago, scientists studying macaques on Koshima Islet found them washing wheat grains free of sand by dropping handfuls in the sea. The wheat floated, while the sand sank. Then the macaques learned that the sea would wash sweet potatoes free of soil. Finally, a young macaque found that swimming and diving were fun. Soon more young macaques were swimming – something they had not done before. The older macaques would have nothing to do with it.

These were new tricks, tried by one monkey and copied by others until they became general behavior. Other tricks of behavior include titmice raiding milk bottles, polar bears raiding cabins, squirrels raiding birdfeeders, and the gulls' much older skill of cracking crabs – probably started in similar ways.

Raiding milk bottles

During a period in Britain when doorstep-delivery milk bottles were fitted with aluminum caps, chickadees in several cities developed a new trick. They found they could sit on the rim of the bottle, peck through the cap, and drink the cream – before the homeowners woke up and took the bottles in.

Cabin raiders

Polar bears have stalked the Arctic for thousands of year, feeding on seals on the sea ice, and on leaves, berries, and bird eggs on land. Trappers and explorers who spend summers in the Arctic must build their cabins strong. If a hungry bear smells food in a cabin, it will break down windows, doors, and walls to get in.

Right: Japanese macaques on Koshima Islet improved their food by washing it in sea water. Young ones watch and copy their elders.

Animals at play

Catching feathers

Adult frigate birds swoop low over tropical seas, watching for flying fish and catching them in flight. Catching a slippery fish in midair demands fast and accurate flying. How do they learn it? Young frigate birds just out of the nest spend hours every day in flight, learning to take off and land in strong winds, to swoop low over the waves, and – like these two – to chase and catch feathers in midair.

Playing mothers and fathers

Adult male fur seals round up pregnant females as they come from the sea, guarding them through their pregnancy. Immature three-to four-year-old males play at rounding up younger seals and herding them together – until the youngsters break out and race off to play on their own.

Do animals play for fun? The truth is most play involves practice for more serious behavior later in life. We are closest to pets when we play with them.

Chasing a ball, running and jumping, or tugging a towel, they are clearly having fun just as we are. We are tapping into an important part of their lives, because this is the way that many higher animals – particularly the social animals – learn how to live.

These fox cubs (opposite page) are just playing, but it is play with a purpose. They are strengthening their muscles and, perhaps more important, they are learning how to bite gently without getting bitten, how to win a round, how to lose and forget a lost battle, how to get along with the family – all good training for one day managing a family of their own. It is a painless way of learning important lessons. Young frigate birds, gentoo penguins, elephant seals, and fur seals also learn by playing during their first months of life.

Play-nesting

Young gentoo penguins raise their first chicks in their third, fourth, or fifth years. In earlier years, they "play-nest." Two-year-olds joining a colony for the first time build rough nests of stones and moss, just like grown-ups, but give up and go swim. A week later they may try again, this time finding a mate, but it's a teenage romance. Next year they'll build a better nest and court more seriously. A year later they might start to breed.

Swimming practice

Elephant seals, huge and ungainly on land, are sleek and fast in the sea. Among the world's strongest swimmers and deepest divers. Their pups spend their first three months on land. Too fat and chubby to swim in the surf, they wallow, blow bubbles, and practice their swimming strokes in safe, sheltered tide pools.

Right: Young foxes spend hours close to their den in play-fighting, a useful training for skills they will need in later life.

Cleaning up

Brushing and combing
Mammals instinctively take care of their fur. This domestic cat, no less than its wild cousins, spends an hour or more every day just licking and grooming. The rough tongue combs the fur into order, and cleans and polishes the claws. A healthy cat cleans regularly. One that neglects its grooming is probably ill.

Do animals groom themselves because, like humans, they like to keep clean and tidy? The truth is that skin, fur, and feathers need to be groomed to keep them weatherproof and prevent build-up of parasites. If the skin is damaged, diseases may enter and other troubles follow. These solemn-looking mandrills (opposite page) are doing what many monkeys and apes seem to enjoy. One is preening the other, combing through the fur with its fingers, pulling out seeds and bits of bark, and finding and eating fleas and ticks. Soon the one that is grooming will, in turn, be groomed. Mutual grooming – doing this for each other – is soothing to both animals. They come to know and trust each other. On a larger scale, it probably helps to hold social groups together.

Other animals, like the cat, heron, and housefly shown here, preen themselves for the simple purpose of protecting their outer defense, while the barber fish cleans others for its own good reasons.

Mandrills, like many other monkeys and apes, spend much of their spare time in mutual grooming – examining and cleaning each other's fur.

Flies and grooming
Houseflies breed in foul, unpleasant places, and carry diseases. Yet they also groom. A fly at rest rubs its forelegs and hind legs together, passing its forelegs over its neck, and its hind legs over its wings. Does that make it cleaner? Not likely, but it probably feels better.

Cleaning and oiling
Birds have a strong instinct to preen their feathers. This makes good sense, because feathers protect the bird's skin, keep it warm, and keep its skin dry if, like the gray heron, it lives or feeds in water. They are the means by which it flies. Running the bill through the feathers brings displaced ones back into position, mends any that are split, and spreads a film of oil (from a gland over the tail) to keep them waterproof.

Barber fish
This small fish swimming dangerously close to those big teeth is a barber fish, so called because it cleans the skin of other fish. Fish sometimes become infested with lice and other skin parasites.

Barber fish make a living from eating the parasites and loose skin, and generally cleaning up the infested fish. For no extra charge, they'll clean inside the mouth and gills, too.

Using tools

Extending the bill

This woodpecker finch lives on the Galapagos Islands, off the west coast of South America. It feeds on grubs, which it digs from holes in the branches of trees. The bird cannot dig the grubs out with its bill. Instead, it breaks off long thorns and uses them to probe out its prey.

Cracking an egg

Here is an Egyptian vulture that has found an ostrich egg. It knows from experience that eggs are good to eat, but a thick, tough shell makes this a hard egg to open. The vulture's bill is not powerful enough to break it. But the bird has learned that a stone will do the trick. It drops the stone repeatedly on the egg, which will sooner or later give way.

Do animals use tools? We used to say, "Man is a tool-using animal," as if man were the only animal that ever used tools. Then we found many different kinds of animals using tools – sticks, stones, or other implements – to do something they could not do otherwise. Tool-using often starts with trial and error. Eggs, crabs, and other breakables crack if you drop them. A twig or stick extends the use of fingers or bill.

The sea otter (opposite page) has discovered how to use two tools at once. It wants to stay still near the surface while feeding, and wants to break into that spiny sea urchin it has scooped from the seabed. To stay still, it wraps itself in loose tendrils

Nutcracker

Chimpanzees and other apes are, of all animals, the closest to humans. Their behavior is, in many ways, similar to our own. They do not think things through as we do and are more easily distracted, but they often use their hands in ways that we recognize. This chimpanzee (right) has learned that nut kernels are good to eat, and a stone will crack the shell.

of kelp – a greenish-brown seaweed that extends from the seabed to the surface. To break into the sea urchin, it carries a stone. In a moment, it will be lying face-upward on the ocean surface, held in place by the kelp, with the sea urchin against its chest, using the stone to break the hard outer casing and feed on the contents.

This sea otter, wrapped in kelp to steady itself at the surface, has brought up a sea urchin and stone from the seabed. It will use the stone to crack open the urchin.

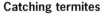

Catching termites

This chimpanzee (left) has learned that a little mound of earth contains lots of tasty termites. It has tried putting its fingers in the hole, but they are too short. Now it is probing with a twig. The termites will bite the twig and hang on, letting themselves be drawn out. Then the chimpanzee will lick them off. Young chimpanzees learn this trick from watching their mothers.

Let's pretend

Feigning injury

Ringed plovers are shorebirds that gather in huge flocks to feed on sandy beaches. They nest on rocks or sand near inland waters. If you approach the nest of a little ringed plover, the parent may run off, staggering, calling, and trailing one wing. Don't be fooled, but move back the way you came. You are probably close to its nest. Dogs, foxes, and other ground predators are attracted to the bird, which leads them away and then flies off, leaving the nest safe.

Feigning death

Honey-eaters are birds that live in Australia, New Guinea, and some Pacific islands. They feed on insects and fruit as well as nectar. If you come near this Australian white-eared honey-eater on its nest, it will stand up, flutter, and seem to drop dead at your feet. Pick it up, and you'll find it completely limp. Even its heart seems to have stopped beating. It will recover soon. This is just another way of diverting attention from the nest and eggs.

Do animals pretend to be dead? "Playing possum" means pretending to die, or keeping out of the way to avoid harm. Long ago, hunters noticed that, if the little animal called the opossum (opposite page) is scared, perhaps by a dog or a fox, it goes limp and falls to the ground as though dead. If left for a few minutes, it comes back to life and scampers off. Pretending to die seems a dangerous thing to do. Wouldn't it pay the opossum just to run? Not if the dog is standing over it. If it runs, the dog will chase after it. If it lies still, it no longer attracts the dog's attention. The dog may still be in a hunting mood rather than an eating mood, and it may wander off to find something else to chase. That gives the opossum a chance to sneak away.

Double bluff

Here is a snake with a double bluff. It is an American hog-nosed snake, so called for its upturned nose. If you come upon it suddenly, it may hiss and try to strike. It has teeth but no poison glands. Then it turns on its back, sticks out its tongue, and feigns death.

Several other animals pretend, or feign, in this way. Do they know what they are doing? No – they are in a kind of shock, and cannot do anything except lie still, which is probably the best thing to do.

Right: Lying on the woodland floor, the American opossum plays dead.

A warning color

Several kinds of frogs play possum, though with a slightly different twist. Frighten this colorful firebelly toad, and it will rise up and topple over backward, showing what you see now – its brilliant underside. It is poisonous, and the vivid color is a reminder. If you have touched one before, it will remind you not to touch it again.

Migration

Some kinds of birds travel thousands of miles each year from one part of the world to another. Like vacationers, they do it to avoid harsh winters. However, they don't travel for fun. This is called migration, and it is a serious business. The birds travel to find places where they can feed better and breed more safely. Are they skilled navigators? Yes, but the methods they use are basically instinctive.

These snow geese have spent the winter in the southern United States, where they have fed in estuaries and damp pastures. Now they are fattened up for their long flight and are heading north, thousands upon thousands of them, navigating by the sun and stars, to breed in Alaska. It's cold there, but their

feathers and fat will keep them warm. Once the snow has gone, there will be plenty of food. They will mate, build nests, lay eggs, and hatch chicks, all during the brief Arctic summer. In August and September they'll return south, to spend another winter in the warmth.

Birds are not the only animals to migrate. Other migrations are made, for example, by bats, butterflies, and locusts. Land migrators include caribou, and sea migrators include whales, fish, and crustaceans.

Tiny traveler
The rufous humming bird, no bigger than a thumb, flies each year from breeding grounds in Alaska and British Columbia to wintering grounds in Mexico. It doubles its body weight with fat – fuel for the journey – before starting out, and flies at heights of up to 12,000 feet (3,650 m).

A long walk
Birds fly, but mammals have to walk. The caribou walks with thousands of others from the northern slopes of Alaska to the forests of northern Alberta and back each year. Calves born in the far northern summer are just big enough to walk south in the fall.

Frequent fliers
Among the world's most northerly breeding birds, turnstones spend northern summers in Alaska, northern Canada, Scandinavia, and Siberia, and winter far to the south, some as far south as Australia, New Zealand, and South America.

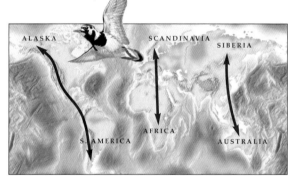

Right: Snow geese take off together in huge flocks, flying in V-formation for thousands of miles to their breeding grounds. Their routes are shown in the inset.

Summer visitors
European swallows catch insects on the fly. Each spring they fly from southern Africa to Europe, to breed as far north as Norway and Sweden. In late summer, they and their fledglings fly back to Africa, a nonstop journey of two or three days.

ALASKA   BAFFIN IS.

N. AMERICA

TEXAS

Index